Nadia Musaninkindi
Jean Jacques Mbonigaba Muhinda
Benson Mochoge

Effect of Cattle manure, mineral fertilizer and rhizobium inoculation

AF153987

Nadia Musaninkindi
Jean Jacques Mbonigaba Muhinda
Benson Mochoge

Effect of Cattle manure, mineral fertilizer and rhizobium inoculation

ISFM and Climbing bean production in Rwanda

LAP LAMBERT Academic Publishing

Impressum / Imprint
Bibliografische Information der Deutschen Nationalbibliothek: Die Deutsche Nationalbibliothek verzeichnet diese Publikation in der Deutschen Nationalbibliografie; detaillierte bibliografische Daten sind im Internet über http://dnb.d-nb.de abrufbar.
Alle in diesem Buch genannten Marken und Produktnamen unterliegen warenzeichen-, marken- oder patentrechtlichem Schutz bzw. sind Warenzeichen oder eingetragene Warenzeichen der jeweiligen Inhaber. Die Wiedergabe von Marken, Produktnamen, Gebrauchsnamen, Handelsnamen, Warenbezeichnungen u.s.w. in diesem Werk berechtigt auch ohne besondere Kennzeichnung nicht zu der Annahme, dass solche Namen im Sinne der Warenzeichen- und Markenschutzgesetzgebung als frei zu betrachten wären und daher von jedermann benutzt werden dürften.

Bibliographic information published by the Deutsche Nationalbibliothek: The Deutsche Nationalbibliothek lists this publication in the Deutsche Nationalbibliografie; detailed bibliographic data are available in the Internet at http://dnb.d-nb.de.
Any brand names and product names mentioned in this book are subject to trademark, brand or patent protection and are trademarks or registered trademarks of their respective holders. The use of brand names, product names, common names, trade names, product descriptions etc. even without a particular marking in this works is in no way to be construed to mean that such names may be regarded as unrestricted in respect of trademark and brand protection legislation and could thus be used by anyone.

Coverbild / Cover image: www.ingimage.com

Verlag / Publisher:
LAP LAMBERT Academic Publishing
ist ein Imprint der / is a trademark of
OmniScriptum GmbH & Co. KG
Heinrich-Böcking-Str. 6-8, 66121 Saarbrücken, Deutschland / Germany
Email: info@lap-publishing.com

Herstellung: siehe letzte Seite /
Printed at: see last page
ISBN: 978-3-659-58862-4

Dedication

I dedicate this work to my family and friends for their love, encouragement and support.

Acknowledgements

I deeply thank God for every blessing in my life, which enabled me to accomplish this research work. I immensely thank my supervisors Prof. Benson Mochoge and Prof. Jean Jacques Mbonigaba Muhinda for their immeasurable professional guidance and enormous efforts provided during this work. I thank Dr Benjamin O. Danga for his assistance in getting started with the research proposal, many thanks to Dr Isaac M. Osuga for his smart orientation in statistical skills and I am sincerely grateful to all Kenyatta University staff for their contribution to my academic training.

I express my gratitude to The Alliance for Green Revolution in Africa (AGRA) for financial and professional support provided during my studies at Kenyatta University. I recognize the collaboration of RAB as well as Mr Chiragaga Dieudonné, NUR Soil laboratory for their valued support.

Family members, colleagues, any person having contributed to the success of this work in one way or another, I say to all thank you.

TABLE OF CONTENT

LIST OF TABLES

LIST OF FIGURES

ABBREVIATIONS AND ACRONYMS

ASHC: Africa Soil Health Consortium

ANOVA: Analysis of Variance

BNF: Biological Nitrogen Fixation

CEC: Cation exchange capacity

cm: Centimeter

CRD: Completely Randomized Design

CV: Coefficient of variation

DAP: Diammonium phosphate

FAO: Food and Agriculture Organization

FEWS NET: Famine Early Warning Systems Network

FYM: Farmyard manure

g: Gram

Ha: Hectare

I: Inoculum

IFIA: International Fertilizer Industry Association

IFDC: International Fertilizer Development Center

ISAR: Institut des Sciences Agronomiques du Rwanda

ISFM: Integrated Soil Fertility Management

kg: Kilogram

LSD: Least Significant Difference

Meq: Milliequivalent

mg: Milligram

mm: Millimeter

MINAGRI: Ministry of Agriculture and Animal Resources

MINECOFIN: Ministry of Finance and Economic Planning

N: Nitrogen

pH: Potential or Power of Hydrogen ions

RAB: Rwanda Agriculture Board

SAS: Statistical Analysis System

SSA: Sub-Saharan Africa

STM: Short Term Mean

USGS: United States Geological Survey

WFP: World Food Programme

ABSTRACT

Agriculture is the major engine of Rwandese economy, accounting for about 40% of the GDP, 85% of employment and 80% of exports. Known as "meat for the poor", beans constitute a predominant source of proteins in Rwandese diet since they supply 65% of national dietary proteins compared to 4% from animal sources. However, the on-farm bean productivity is about $0.8 - 1.0$ tons/ hectare which is quite low compared to 5 tons/hectare that is achieved under optimal management conditions. The aim of this study was therefore to determine the effect of cattle manure, mineral fertilizer and Rhizobium inoculation on production of climbing beans and subsequently the soil properties in Burera District. The experimental design was a split plot in completely randomized design (CRD) with two main plots (with and without Rhizobium inoculum); four sub-plots (Cattle Manure, DAP, Cattle manure + DAP, untreated control) with quantities applied at single level for each treatment, i.e. 20t/ha for Cattle manure, 50 kg/ha for DAP and 100 g of inoculum which was mixed with 15 kg of beans. The experiment involved 8 treatments which were replicated three times to give 24 plots. The mean bean grain yields from inoculated treatments and non-inoculated treatments showed statistically significant difference (P< 0.0001), that is 3900 kg/ha from inoculated plots and 2946 kg/ha from non-inoculated. Statistical significant differences were also found among treatments (P<0.0001) with the highest mean yield of 4782 kg/ha obtained from treatment Inoculum + DAP + Cattle Manure against 2640 kg/ha from untreated (control) plots. The mean number of nodules was significantly different (P< 0.0001) between inoculated (60 nodules) and non-inoculated (15 nodules) plots. The highest number of nodules (95) was recorded from plots that were treated with Inoculum + DAP + Cattle Manure and the lowest (14) in the untreated control plots (P<0.0001). Regression analysis between yield and nodule number showed a coefficient of determination R^2 of 0.8 and a p value of < 0.0001, which confirmed the dependence of the yield on nodules number. In terms of cost-benefit analysis, in the highest yielding treatment (I+FYM+DAP) scenario, a farmer is likely to earn around 1,330 USD per season per hectare; while in the middle and lowest yielding treatment (I and UNTREATED CONTROL), the farmer is likely to lose 43.8 USD and 388 USD per season per hectare, respectively. On the effect of treatments on soil chemical properties, no tangible changes were observed in pH, CEC and organic matter at the end of season. According to these results, a combination of mineral fertilizer, inoculum and cattle manure application gave the best results in terms of bean yield, nodulation and nitrogen uptake and therefore could be better considered for recommendation to climbing bean growers in the region.

CHAPTER ONE

INTRODUCTION

1.1 General background

Rwanda is a landlocked country with a total area of 26,338 km^2. According to the estimates of 2002 census, Rwanda population is 10 million people and it is estimated to be 15 millions by the year 2020. With 310 inhabitants per km^2, Rwanda is among the countries of Africa with the highest population density and the majority live in rural areas. The Rwandan climate is conditioned by landscape in such a way that the farther to the west, the lower the altitude, the warmer the temperature and the lower the precipitation with annual rainfall ranging between 900 and 1,600 mm (MINAGRI, 2004).

Agriculture is the major engine of Rwandan economy, representing about 40% of the GDP and accounts for approximately 85% of employment and 80% of exports. Agriculture is mainly rainfed, with mixed cropping being a dominant farming system (MINECOFIN, 2002).

Known as "meat for the poor", beans are regarded as a "near complete food" based on their nutritional values. A part from proteins, they provide resistant starch, soluble and insoluble fiber, vitamin B and minerals such as Iron, Zinc, Magnesium, Copper and Potassium (Hutchins, 2011) and they account for 65% of national dietary proteins compared to 4% from animal sources. Research has shown that bean consumption in Rwanda is the highest in the world, where annual average for Africa is evaluated at 17 kg per capita compared 60 kg per capita in Rwanda (WFP, 2009). Climbing beans are more profitable given that they save land by assuring higher productivity on the same area compared to bush beans (3:1) (Kelly *et al.*, 2012) and low disease incidence on pods far from the soil surface. On-farm productivity of about 0.8 – 1.0 t ha^{-1} is still low for climbing beans compared to 5 t ha^{-1} that are achieved with optimal management conditions (ISAR, 2009). Diseases, lack of improved seeds, small size plots, low use of fertilizers, and lack of cheaper innovations of staking are the main reasons for low productivity experienced by farmers. However, the Government of Rwanda and partners in development are putting more efforts towards the development of climbing beans' program. In that context, climbing bean is among the selected crops in Crop Intensification Program, which is more profitable in terms of accessing inputs and value addition to small land holdings. This study therefore seeks to contribute to more knowledge about climbing beans production using simple farmer friendly nutrient management alternatives.

1.2 Problem statement

Agriculture in Rwanda remains unproductive due to intensive exploitation of shrinking land brought about mainly by high population density. In that context, 56% of farm households exploit less than 0.5 hectare and with no simultaneous measures to maintain soil health; which results in soil fertility decline (MINAGRI, 2004). Though expensive to small scale farmers, mineral fertilizers are still not efficiently used in terms of rates, time of application as well as choice of the right fertilizer. Cattle manure, as one of farm available organic resources is not well managed in order to transfer its benefits to soils (Uphoff *et al.*, 2006; Azeez and Van Averbeke, 2010). Moreover, the combination of mineral and organic fertilizers is not well understood by smallholder farmers and yet it results in increased yields and good soil conditions (Sanchez, 2002). This study seeks to evaluate the effect of inoculation, organic and inorganic fertilizers application on climbing beans yield as well as soil properties.

1.3 Objectives

The main objective of this study was to contribute to climbing beans knowledge base through application of cattle manure, rhizobium inoculation and mineral fertilizer on climbing beans production and soil properties changes in Burera District.

The specific objectives to achieve this main objective were:

1. To determine the effect of cattle manure, DAP and Rhizobium application on the yield of climbing beans;
2. To determine the effect of cattle manure, DAP and Rhizobium application on Climbing beans nodulation
3. To determine the effect of cattle manure, Diammonium phosphate (DAP) and Rhizobium inoculation on soil chemical properties under climbing beans;
4. To carry out cost-benefit analysis of different nutrient management strategies in climbing bean production in Burera District.

1.4   Research hypotheses

1. Cattle manure, DAP and Rhizobium inoculation have a significant effect on climbing beans yield;
2. Cattle manure, DAP and Rhizobium inoculation have a significant effect on climbing beans nodulation.
3. Cattle manure, DAP and Rhizobium inoculation have a significant effect on soil chemical
       properties;
4. There is a net positive return when a farmer applies the highest yielding treatment.

1.5 Justification of the study

The two main strategies for improving nutrient availability in cropped ecosystems are to increase inputs and to reduce losses. Improving a cropping system's nutrient-use efficiency requires matching soil nutrient release – whether from organic or inorganic sources – with the demand for nutrients by plants. For farmers who have manure available, it is an opportunity to derive benefits from the combined application of manure and fertilizer to crops. In ISFM context, cattle manure is not only a safe and effective way of recovery for lost plant nutrients like nitrogen and phosphorus, but also improves the physical and chemical attributes of the soil. Climbing beans show a high potential in alleviating protein malnutrition and poverty among rural population given their higher productivity compared to bush beans. Cattle manure is best provided by the zero grazing approach which is currently being promoted in the whole country. By complementing this to mineral fertilization which is being promoted by Crop Intensification Program (CIP), yields are improving but still low (MINAGRI, 2011). Furthermore, inoculation in bean production is still less adopted by farmers, which needs more advocacy and research works so that yields can be substantially increased, thus improving smallholder farmers' livelihoods through food and income availability. With regards to application of mineral fertilizer, cattle manure and rhizobium inoculation on Climbing beans, the Rwanda Agriculture Board (RAB) conducted pilot experiments on farmers' fields and hence this research work (in a farmer-researcher managed experiment) was carried out to provide more information.

1.6 Conceptual framework

Relatively low fertility of soils under this study is manifested in chemical properties, which in turn result in low yields (Table 1). By adding cattle manure combined with inorganic fertilizer (DAP) as well as Rhizobium inoculation, we expect soil physical, chemical and biological properties to be improved and hence lead to increased yields of climbing beans. Soil properties such as pH, CEC, nitrogen and organic matter are to be measured to evaluate dynamic nutrient changes in soil, and hence plant nutrients uptake and crop yields.

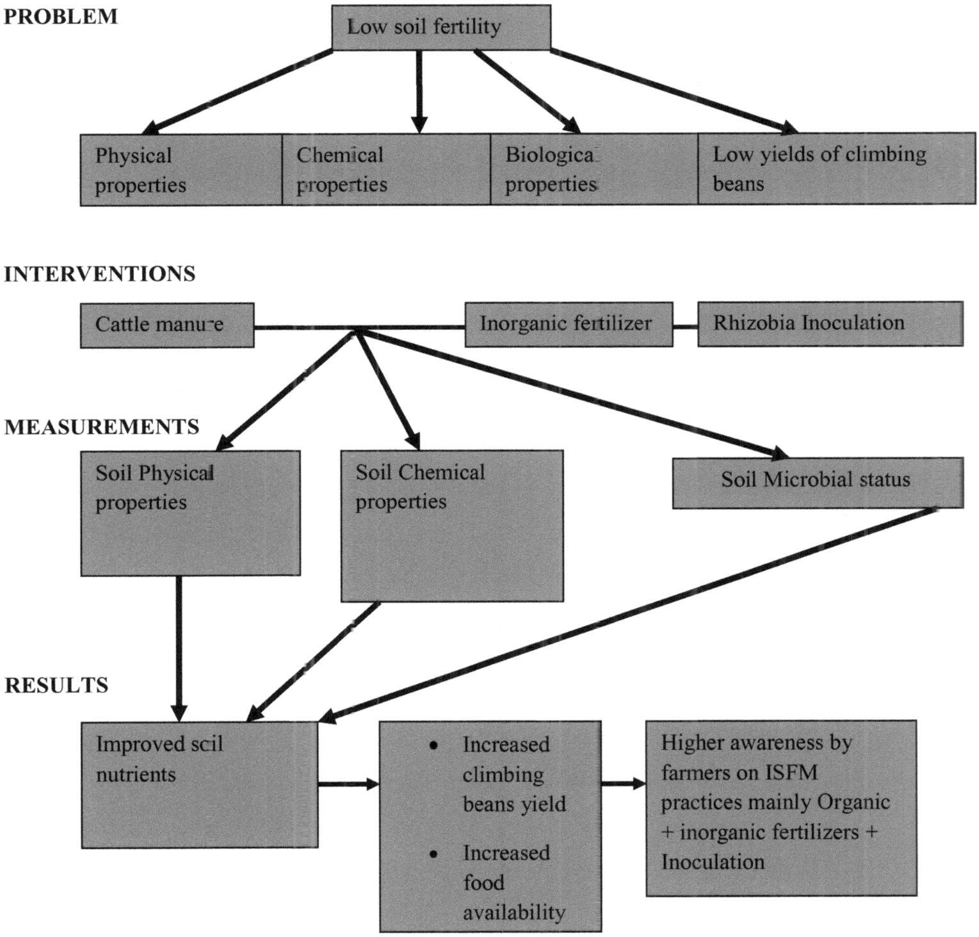

PROBLEM

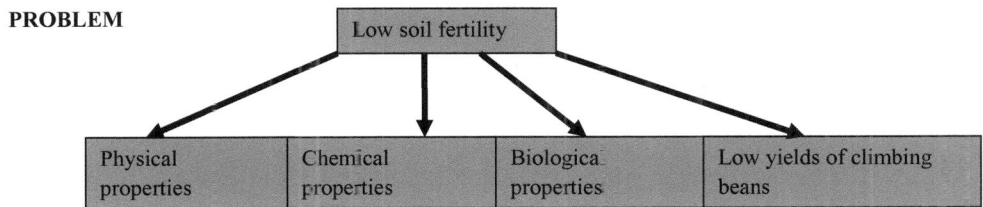

```
                        ┌─────────────────────┐
                        │   Low soil fertility │
                        └─────────────────────┘
```

| Physical properties | Chemical properties | Biologica_ properties | Low yields of climbing beans |

INTERVENTIONS

| Cattle manure | | Inorganic fertilizer | Rhizobia Inoculation |

MEASUREMENTS

| Soil Physical properties | Soil Chemical properties | Soil Microbial status |

RESULTS

| Improved soil nutrients | • Increased climbing beans yield • Increased food availability | Higher awareness by farmers on ISFM practices mainly Organic + inorganic fertilizers + Inoculation |

Figure 1: The Conceptual framework ɔn the production of climb_ng beans

CHAPTER TWO

LITERATURE REVIEW

2.1 Overview

Reasons for declining crop yields in Sub-Saharan Africa (SSA) are multiple and include frequent and prolonged droughts, low quality seeds with low yield potentials; poor crop husbandry practices such as poor seedbed preparation, late planting along with poor spacing as well as poor weeding and non-control of diseases and pests. Moreover, poor and unstable economies and policies related to acquisition of inputs, value addition and good markets for products; post harvest losses of produce due to inadequate preservation, storage and transportation; declining soil fertility resulting from continuous cultivation of high nutrient demanding crops such as maize on the same pieces of land without or with minimal nutrients returns (Okalebo, 2009; Okalebo *et al.*, 2004) have immensely contributed to the poor performance of crop yields.

Poor soil fertility and nutrient depletion continue to pose huge obstacles to securing the needed good harvests and sustainable soil productivity. This, therefore, imperatively calls for better soil fertility management approaches in countries of SSA (Sanginga and Woomer, 2009; Mafongoya *et al.*, 2006). Synthetic fertilizers and organic inputs in most cases fulfill different functions in soil fertility restoration. Given that neither of them is widely available to smallholder farmers; it makes more practical sense to advocate for their combined use. One of the benefits of integrated use of input resources is a direct result of more nutrients being added from the two or more resources combined than applied singly. Synthetic fertilizers have the advantage of being readily soluble in soil solution, less bulky and easy to manipulate but their constitution in most cases does not include the much needed essential minor elements as compared to organic manures which meet this requirement (Bekunda *et al.*, 2010).

In terms of improving soil structure, the active and some of the resistant soil organic components, together with micro-organisms (especially fungi), are involved in binding soil particles into larger aggregates. Aggregation is important for good soil structure, aeration, water infiltration and resistance to erosion and crusting (Bot and Benites, 2005).

Furthermore, the integration of legumes in African Integrated Soil Fertility Management (ISFM) practices offers great importance given the numerous benefits it provides. Many field legumes produce high yielding grains that greatly improve household diets; they provide livestock feeds and their crop residues offer benefits to soil fertility through biological nitrogen fixation that, in turn reduce the requirement for costly mineral fertilizers (FTF, 2011). Small-scale farming household that has incorporated legumes into its enterprises is in a better position to raise its wellbeing and to meet expectations in improved living standards (Sanginga and Woomer, 2009).

2.2 Effect of mineral fertilizers on crop production and their availability in Rwanda

Mineral fertilizers are used to supplement the natural soil nutrient supply in order to satisfy the demand of crops with a high yield potential and produce economically viable yields; compensate for the nutrients lost by the removal of plant products, by leaching or gaseous loss (IFIA, 2000). In Rwanda, mineral fertilizers have contributed to increased yields of climbing beans even though achieving potential yields is still a challenge due to other important factors involved in crop production (MINAGRI, 2011). Most recent projective studies of global agricultural production into the 21st century, suggest that a global food crisis is unlikely but that many countries and regions will continue to suffer from chronic malnutrition. From a resource perspective, growing world population and per capita incomes will likely require more intensive agricultural crop production. Higher yields will in turn increase the demand for agricultural inputs, especially mineral fertilizers (FAO, 2000). Lack of credit, poor marketing capabilities, high transport costs, lack of availability of fertilizer, inadequate demand to stimulate investment in production and distribution, lack of crop markets, devaluation of domestic currencies and weak extension services constrain fertilizer use. The lack of credit has been identified as a major determinant of fertilizer use in many African countries including Rwanda especially for poor and middle households (Wallace and Knausenberger, 1997). The assessment of agricultural inputs market in Rwanda reveals that agri-inputs use in Rwanda is among the lowest in Africa, only 15% of the farmers used mineral fertilizers in 2005 while 26% of the farmers in 2011 were reported having used mineral fertilizers alone or in mix with organic fertilizers (IFDC, 2009; MINAGRI, 2011).

The reasons are mainly unavailability of right agri-inputs at the right time and of right amounts in rural areas. Equally important is the cost of agri-inputs, farmers' purchasing power, poor extension services and the weakness of the private sector. Fertilizer recommendations are outdated; hence the private sector does not stand confident (IFDC, 2009). However, many reports in the literature have showed that continuous use of sole fertilizers may lead to shortage of nutrients not supplied by the chemical fertilizers and may also lead to chemical soil degradation (Mafongoya *et al.*, 2006).

2.3 Importance of combined organic and mineral fertilizers in soil health

The goal of ISFM is to maximize the interactions that result from the potent combination of fertilizers, organic inputs, improved germplasm, farmer knowledge and adaptations to local conditions and seasonal events (Sanginga and Woomer, 2009). The ultimate outcome is improved productivity, enhanced soil quality, and a more sustainable system through wiser farm investments and field practices with consequent minimal impacts of increased inputs use on the environment. Combining mineral and organic inputs results in greater benefits than either input alone through positive interactions on soil biological, chemical and physical properties (Alley and Vanlauwe, 2009; Bekunda et al., 2010).

The combined application of manure and fertilizers may be attractive for responsive fields in the short term and maintain soil Carbon in the long term (IFIA, 2000). In that context, it is crucial to note that greater crop productivity induced by the use of mineral fertilizers does not translate into better soil fertility in the long term when large amounts of Carbon and nutrients are removed every season from the fields with the crop harvests residue (Bekunda et al., 2010).

Although animal manure remains an option to manage soil fertility in mixed smallholder crop-livestock systems, its availability and quality are often poor and moreover, the rates applied and nutrient concentration of manures used are low in most field experiments (Tittonell et al., 2008). In Rwanda, as in other SSA countries, farmers still need to pursue sustainable intensification to maintain food security, mitigate the effects of weather variability and climate change, protect land and increase incomes (IFDC, 2002). ISFM is a sustainable approach that acknowledges the need for both organic and mineral inputs to sustain soil health and crop production due to positive interactions and complementarities between them (ASHC, 2012).

2.4 The role of legumes in ISFM

The use of organic matter or microbial inoculation to increase soil fertility is referred to as Bio-fertilization. Rhizobium plays an important role in agriculture by inducing nitrogen fixing nodules on the roots of legumes (Awad and Eltahir, 2011).

This mechanism is not new in the scientific world yet, until today, very few farmers know the technology. Most leguminosae (about 90%) can establish a symbiotic association with aerobic diazotrophic gram-negative bacteria commonly referred to as rhizobia. Legumes are known to fix nitrogen in the soil, converting nitrogen in their environment into ammonia, a nitrogen-containing compound that they are able to use (Burdass and Hurst, 2002). Participation in legume enterprises by small-scale farmers has numerous benefits, both direct and indirect. Many field legumes produce high yielding grains that greatly improve household diets. Legumes provide livestock feed and their crop residues offer benefits to soil through biological nitrogen fixation that, in turn reduce the requirement for costly mineral fertilizers (Sanginga and Woomer, 2009). However, the challenge of increasing legume enterprises within small-scale farms in Africa has several key aspects relating to biology, crop genetic improvement, useful technologies, rural economic, cultural perspectives and political support (Sanginga and Woomer, 2009).

2.5 Role of Rhizobia inoculation in legumes production

The Rhizobia were first isolated and cultured from nodules of a number of different legume species by Martinus Beijerinck of Holland in 1888. Since then, Rhizobium has been found to consist of the following main genera Bradyrhizobium, Sinorhizobium, Azorhizobium, Mesorhizobium (Hirsch, 2009, Fujishige et al., 2008).

Biological nitrogen fixation is of capital importance and consists in the reduction of molecular nitrogen (N_2) to ammonia (NH_3), providing the Earth's ecosystems with about 200 million tons N per year. It has been estimated that the 80-90% of the nitrogen available to plants in natural ecosystems originates from biological nitrogen fixation (Saikia and Jain, 2007). Nitrogen constitutes almost 80% of the atmosphere, but is metabolically inaccessible to plants due to the exceptional triple covalent bond ($N\equiv N$). The ability to catalyze enzymatic reduction of N_2 to NH_3 is limited to a variety of microorganism defined as nitrogen-fixing or diazotrophic microorganisms which are widely distributed in all ecosystems as either free-living organisms or in symbiotic association with a number of different legume types. These nitrogen fixing prokaryotes can be anaerobic, facultative aerobic, aerobic, photosynthetic or non-photosynthetic. All carry out N_2 reduction by an enzymatic complex termed nitrogenase (Bot and Benites, 2005).

Commonly, rhizobia are found in soils but may not produce effective nodulation mainly because they are too few or need a better legume type. However, they are known for their ability to ascertain a symbiosis with legumes as they infect and inhabit root nodules, where they carry out atmospheric nitrogen fixation and make it available to the plant (FAO, 1984; Sessitsch et al., 2002).

2.6 Effect of soil organic matter on soil properties

Organic matter affects both the chemical and physical properties of the soil and its overall health. Properties influenced by organic matter include: - soil structure; moisture holding capacity; diversity and activity of soil organisms, both those that are beneficial and harmful to crop production; and nutrients availability. It also influences the effects of chemical amendments, fertilizers, pesticides and herbicides (Rascio and La Rocca, 2008).

Organic matter influences the physical conditions of a soil in several ways. Plant residues that cover the soil surface protect the soil from sealing and crusting by raindrop impact, thereby enhancing rain water infiltration and reducing runoff. Surface infiltration depends on a number of factors including aggregation and stability, pore continuity and stability, the existence of cracks, and the soil surface condition (Abiven *et al.*, 2008; Ashman *et al.*, 2008). Increased organic matter contributes indirectly to soil porosity (via increased soil faunal activity).

Organic matter contributes to the stability of soil aggregates and pores through the bonding or adhesion properties of organic materials such as bacterial waste products, organic gels, fungal hyphae and worm secretions and casts (Tang *et al.*, 2011). The chemical and nutritional benefits of organic matter are related to the cycling of plant nutrients and the ability of the soil to supply nutrients for plant growth. Organic matter retains plant nutrients and prevents them from leaching to deeper soil layers.

Microorganisms are responsible for the mineralization and immobilization of Nitrogen, Phosphorus and Sulfur, thus contribute to the gradual and continuous liberation of plant nutrients. CEC is linked closely to the organic matter content of the soil. It increases gradually with time where organic residues are retained, first in the topsoil and later also at greater depth (Rascic and La Rocca, 2008).

CHAPTER THREE

MATERIALS AND METHODS

3.1 Description of the study site

This study was conducted in Burera District, in the highlands of the Northern Province of Rwanda, where the climbing beans are most produced. Other important staple crops in this area include irish potato, maize and sweet potato; all cropped in a relatively favorable bimodal rainfall regime. Burera Disrict population is growing, with the density having recently reached 522 inhabitants /Km^2 (NISR, 2012); which leads to smaller plots available for agriculture production. The District baseline report by NISR (2008) showed that 54% of the population in Burera own land of less than 0.5 ha.

It has a wet climate of two dry seasons (Mid December-February; June-September) and two rainy seasons (March-Mid June; Mid September-December). The annual rainfall in Burera is estimated between 1400 and 1800mm, and the temperatures range from 10 to 25°C. The district has an average altitude of 2,011 meters and is characterized by steeply sloping hills. The soils in the area are Andosols which are mainly influenced by volcanic activity. Our experiment was carried out during season 2012A which normally starts in September and ends in February. The figure below (Figure 2) highlights Burera District location on the map of Rwanda.

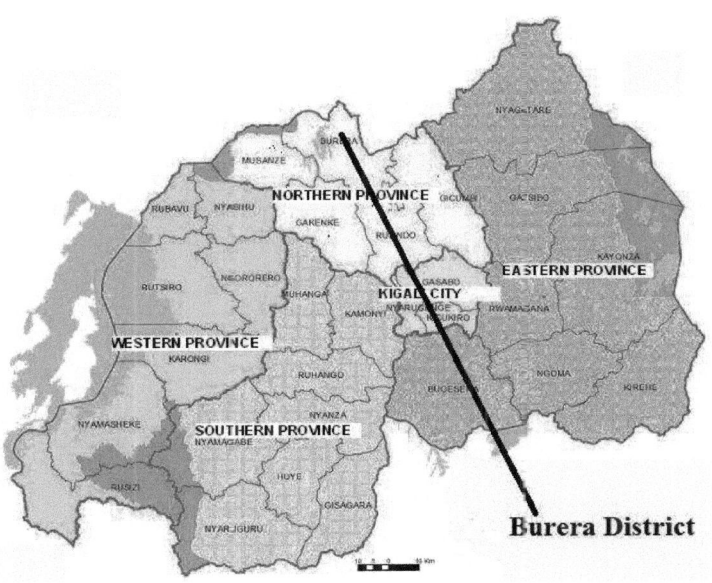

Figure 2: Administrative Map of Rwanda

3.1.1 Initial soil fertility status

The initial soil fertility status (Table 1) analyzed before the commencement of this experiment showed that the soil was fairly fertile in terms of soil physical properties and in some chemical properties except for available nitrogen and phosphorus which were somewhat low. This therefore, calls for some additional inputs (nitrogen and Phosphate fertilizers) to increase climbing bean yields. The soil is moderately acid and good for beans production (Ebesu, 2004; RAB, 2012). The soil structure is sandy loam, which indicates a high infiltration rate due to the fact that it has coarse texture and large porous spaces which promote fast infiltration (Makungo and Odiyo, 2009; Hanson and Orloff, 1998). Such well drained soils have a great leaching potential, which usually affect mobile nutrients such as nitrogen in form of nitrates (Czymmek *et al.*, 2005). The soil's aggregate stability is fairly high, an indication of high ability to resist erosion and therefore, not susceptible to dispersal during rainstorms (Herrick *et al.*, 2001). The soil CEC of the trial site is very high, almost comparable to   organic matter which is also very high (8%), mainly due to the influence of the coldness and wetness of the area (Franzluebbers, 2001). Organic matter has been reported to have a four to 50 times higher CEC per given weight than clay (Shoji *et al.*, 1993; Ketterings *et al.*, 2007).

Table 1: The physical and chemical soil properties of the experimental field at the start of trial in September 2011.

Soil Properties	0-15 cm	15-30 cm
pH water	6.01	5.94
pH KCl	5.27	5.29
OM %	8	8
N-NO$_3$ (mg/kg)	8.29	8.30
N-NH$^+_4$ (mg/kg)	10.63	10.64
Total Nitrogen N (mg/kg)	21.4	21.4
Available P (mg/kg)	18.82	18.89
Potassium K (mg/kg)	74	74
CEC (meq/100g)	48.93	53.73
Exchangeable Al^{3+} (meq/100g)	–	–
Exchangeable H$^+$ (meq/100g)	0.47	0.47
Aggregate stability %	41	41
% clay	14	14
% Silt	32	32
% Sand	54	54
Textural class	Sandy loam	

Source: Our research work

Cattle manure used in the experiment was also analyzed for chemical properties and had 0.5 % of total nitrogen, 0.3 % of phosphorus, 0.5 % of potassium and 0.4 % of sulfate.

19

3.1.2 Rainfall status during the season

As it is shown on   Figure 3, the rainfall during the cropping season (third dekad
of September 2011-third dekad of January 2012), was mostly above average in
the months of September, October and December 2011 but fell below average in
November 2011 and January 2012). The average mentioned here refers to Short
Term Mean (STM) as calculated from 2003 to 2008 rainfall measurements from
Burera meteorological station.

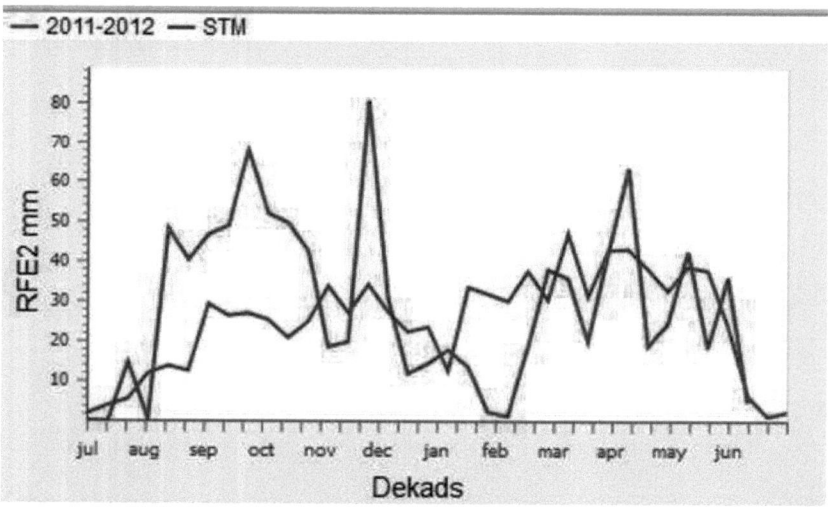

Figure 3: Burera Rainfall Estimate

Source: FEWS NET/Early warning explorer/USGS.

3.2 Field experiment

3.2.1 Experimental field design and treatments

The experiment set up in season 2012 A (September 2011) was a split plot in a completely randomized design (CRD) with two main plots (With and Without Inoculum); four sub-plots (Cattle manure, DAP, Cattle manure + DAP, Control) at two levels for each; i.e 0 and 20 t/ha (Cattle manure) and 0 and 50 kg/ha (DAP). For inoculated plots, 100g of the inoculants was thoroughly mixed with 15 kg of beans before planting. The experiment had eight treatments which were replicated three times. A total of 24 plots were used in the experiment.

3.2.2 Experimental layout (Split plot in CRD)

4m

5m

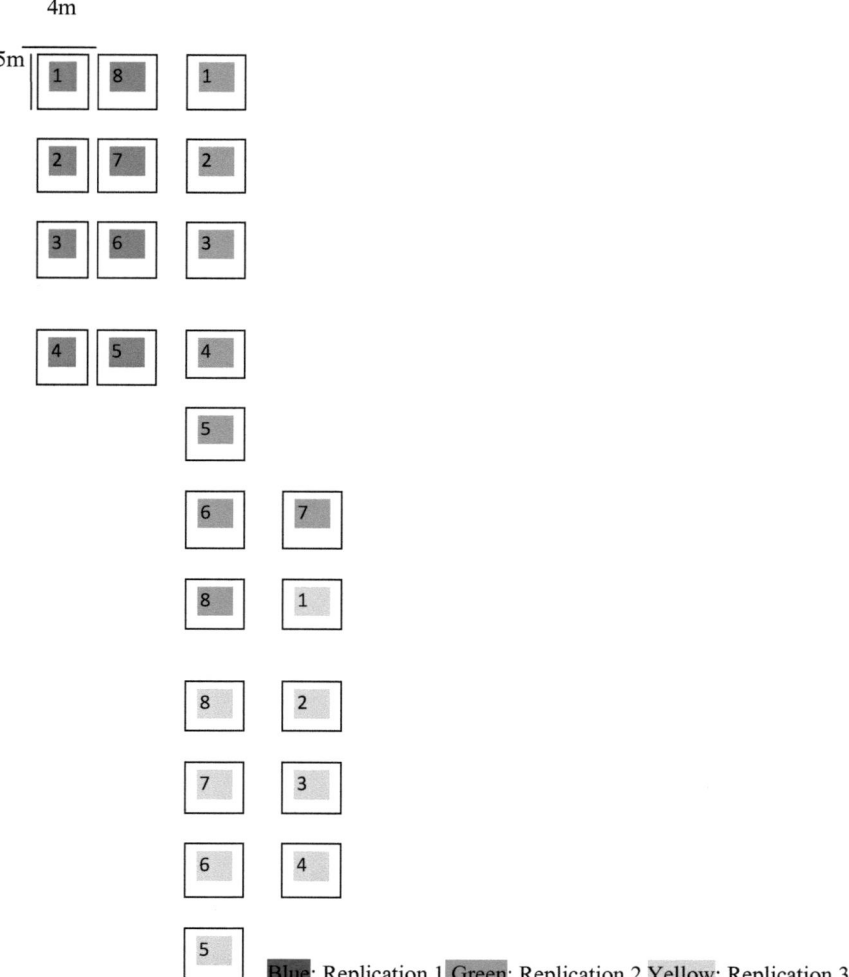

Blue: Replication 1 Green: Replication 2 Yellow: Replication 3

Figure 4: Experimental field layout at Burera during season 2012 A.

The experimental layout was influenced by the terrain of the field which was slightly sloppy towards the east, and the treatments' replications.

3.2.3 Field experimental management

Before the experimental setting, land was tilled by hoe and racked to obtain a fine tilth. Ridges were also put in place so as to limit water moving from one plot to another. Plots sizes were 4×5 m, with 20 cm spacing between plants and 50 cm between rows, which gave 200 plants per plot and 4,800 plants for the whole experiment.

Climbing bean, (Gasilida variety) was planted as a test crop and inoculation process included adding two tablespoonfuls of sugar to 300 ml of clean warm water in a bottle; which was shaken thoroughly to dissolve the sugar. The well dissolved sugar solution was then poured onto the basin containing 15 kg of beans; mixed beans with sugar solution until all the seeds were wet. After that, the content of the BIOFIX packet was poured onto the wet seeds in the basin and mixed thoroughly until all seeds were uniformly covered with the inoculants. The inoculated seeds were then covered to protect them from direct sunlight. Inoculated seeds were immediately planted and to each planting hole, cattle manure and DAP were applied and were thoroughly mixed with the soil in order to avoid adverse effects on the seeds by the inputs.

Weeding was done twice; the first one, two months after planting and the second after three months and half during the growth period. The pests and diseases that were observed in the course of the crop growth period such as bean foliage beetle at early stages of germination, some aphids at flowering and Ascoschyta leaf spot during pod filling phase, were all controlled by applying the appropriate chemicals.

3.3 Soil sampling and analyses.

Soil sampling of the experimental site was done at the beginning and at the end of experiment. The soil samples were obtained across the field at 0-15 cm and 15-30 cm depths using transect method. The soil samples of each depth from six different spots in each plot were taken and mixed to obtain a composite sample that was analyzed and used to characterize the soil's initial fertility status (Table1). The soil pH, soil organic carbon, total nitrogen, available nitrogen (NH_4^+ and NO_3^-), Available P, Potassium, Cation exchange capacity, exchangeable aluminum and hydrogen; texture, and aggregate stability were analyzed in the laboratory.

3.3.1 Soil pH analysis

20 g of soil were weighed in a 60 millilitres (ml) bottle then 50ml of distilled water (for pH water which is active pH) was added with a dispenser. The mixture was stirred for 10 minutes, allowed to stand for 30 minutes (min) and stirred again for 2 min. The pH value was then measured with a pHmeter 30 to 60 seconds until the values remained constant (Okalebo *et al.*, 2002). The electrode was then removed from the bottle, rinsed with distilled water before introducing it to the next sample. Similar procedure was followed in the determination of potential pH (pH KCl) but this time using KCl 1M instead of distilled water.

3.3.2 Soil texture analysis

Bouyoucos or Hydrometer method was used for soil particle size analysis (Okalebo et al., 2002).

A sample of 50 g of air dry <2 mm soil was weighed out into a 400 ml beaker and was saturated with distilled water before adding 10 ml of 10% Calgon solution. The suspension was transferred to the dispersing cup and added about 300 ml of tap water. The suspension was mixed for 2 minutes with an electric high speed stirrer then transferred into a graduated cylinder. The cylinder was covered with a tight-fitting rubber band and the suspension was mixed by inverting the cylinder carefully ten times. The time was noted and 2-3 drops of amyl alcohol were quickly added in order to remove froth and after 20 seconds the hydrometer was gently placed into the column. The hydrometer readings and thermometer measurements at 40 seconds were recorded. The cylinder was covered again with a tight-fitting rubber band and the suspension was mixed by inverting the cylinder ten times and then allowed to stand undisturbed for 2 hours after which both hydrometer and temperature readings were taken. The % sand, silt and clay were then calculated using the formulas in the Appendix 4 (ii).

3.3.3 Soil organic carbon analysis

Walkley and Black (colorimetric method was used (Okalebo et al., 2002). A sample of 0.3g of ground soil (<0.5mm) was weighed out into a clean labelled 100 ml digestion tube, 2 ml of distilled water, 10 ml of potassium dichromate 5% were added and allowed it to completely wet the soil, 5 ml of concentrated sulphuric acid were then added from a slow burette and the mixture was gently swirled to mix. The mixture was digested at 150°C for 30 min and allowed to cool, then added 50 ml of 0.4% barium chloride. It was swirled to mix thoroughly, then brought to 100 ml mark by distilled water and allowed to settle overnight so as to leave a clear supernatant solution. An aliquot of the supernatant solution was transferred into a cuvette, and measurement of absorbance of the standards, the sample and the blank at 600 nm was performed. The content of total organic carbon in air dry soil expressed in % C was then calculated using the formula in the Appendix 4 (iii).

3.3.4 Total nitrogen analysis in plant leaves

Kjeldahl method was used to determine the total Nitrogen in plant leaves
(Okalebo *et al.*, 2002). A sample of 0.3 g of oven dried (70°C) and grounded
plant tissue (<0.25 mm, 60 mesh) was weighed into a dry clean tube and 2.5 ml
of a mixture (Salicylic acid dissolved in sulphuric acid-selenium mixture) was
added to each tube and the reagent blanks for each batch of samples. They were
then digested at 110°C for 1 hour, removed to cool and added three successive 1
ml portions of hydrogen peroxide before raising the temperature to 330°C and
continue to heat. When the solution turned colourless and remaining sand white,
they were allowed to cool and then 25 ml distilled water was added and mixed
until no more sediment dissolved. Total N was then determined in the digests
through distillation whereby free ammonia was liberated from solution by steam
distillation in presence of excess alkali (NaOH). The distillate was collected in a
conical flask containing excess boric acid with drops of mixed indicator.
Titration of the distillate was then carried out using N/70 HCl until colour
changed from green to pink. The total ml of N/70 HCl used were recorded and
used to determine total N using the formula in the Appendix 4 (i).

3.3.5 Soil available nitrogen ($NH_4^+ + NO_3^-$) analysis

Devarda's alloy method (Okalebo *et al.*, 2002; Pauwels *et al.*, 1993) was used.
10 g of soil sample were weighed and kept in refrigerator into a plastic shaking
bottle; added 100 ml of 1 M KCl extracting solution and the content was shaken
for 1 hour. The mixture was filtered through Whatman filter paper No 42. 5 ml
of boric acid indicator solution were added to a 50 ml conical flask having a
calibration of 30 ml and the flask was placed under the condenser of the steam
distillation so that the end or tip of the condenser was about 40 cm above the
surface of the boric acid indicator solution. An aliquot of 10 ml of the soil

extract was pipetted into the distillation flask and added about 0.2 g (scoop) of ignited (and cool) MgO directly to the bulb of the distillation flask. The flask was then attached to the distillation apparatus using spiral steel springs and started distillation by closing the stopcock on the steam-bypass tube. When the distillate reached 30 ml mark on the receiver conical flask, distillation was stopped by opening the stopcock on the steam-by pass tube, rinsed the tip of the condenser with a little distilled water. Ammonium N content in the distillate was determined by titration with 0.01 N sulfuric acid placed in a micro burette. The color change at the end point was from green to a permanent faint pink. After distilling Ammonium N, from the sample extract in the above procedure, we removed the stopper from the side arm of the distilling flask, added 0.2 g of Devarda's alloy using a dry powder funnel to reach the bulb of the flask; replaced the stopper immediately. We distilled nitrates in fresh boric acid, the nitrate is converted to ammonium and trapped in the conical flask, and this ammonium was then estimated by titration with 0.002 N sulfuric acid as described above. The formulas used to calculate nitrates and ammonium are shown in Appendix 4 (iv).

3.3.6 Soil available phosphorus analysis

The method used was Bray 1 (Pauwels *et al.*, 1993); 2.5 g of soil were weighed into a 250 ml plastic bottle, added 25 ml of the Bray P1 extracting solution (NH_4F: ammonium fluoride) and was shaken for 5 minutes. Extracts were filtered using Whatman No. 42 filter paper into clean 50 ml bottles. Phosphorus was then analyzed by colorimetry using a blank and standards prepared in the Bray P-1 extracting solution (Bray and Kurtz, 1945). The formula is presented in the Appendix 4 (v).

3.3.7 Cation Exchange Capacity

A sample of 25.0 g of soil was weighed into a 500 ml Erlenmeyer flask and to it was added 125 ml of the 1 M NH4OAc, then shaken thoroughly, and allowed to stand for 16 hours (or overnight). A 5.5 cm buchner funnel with retentive filter paper was fitted, the paper was moisten, applied light suction, and the soil was transferred. The soil was gently washed 4 times with 25 ml of NH_4OAc, allowing each addition to filter through but not allowing the soil to crack or dry. The soil was washed with 8 separate additions of 95% ethanol to remove excess saturating solution and each addition was allowed to filter through before adding more. The adsorbed NH_4 was extracted by leaching the soil with 8 separate 25 ml additions of 1 M KCl, leaching slowly and completely as above. Soil was discarded and the leachate transferred to a 250 ml volumetric flask. The concentration of NH_4-N in the KCl extract was then determined by distillation. Also NH_4-N in the original KCl extracting solution (blank) was determined to adjust for possible NH_4-N contamination in this reagent (Chapman, 1965). The CEC in meq/100g was then obtained using the calculations in Appendix 4 (vii).

3.3.8 Aggregate stability analysis

Soil aggregate stability was determined using wet sieving method (Kemper and Koch, 1966; Carter and Gregorich, 2006). 4 g of soil samples, 2mm air-dried were pre-moistened, placed in the sieves (0.25 mm), the latter were immersed in the cans filled with distilled water, were then placed in the tackle box which was covered and moved up and down at the rate of 34 cycles/minute for three minutes through a vertical distance of about 1.3 cm. After wet sieving, the sieve holder was raised out of the water and placed in the leak out position. When there was no water leaking out of the sieves anymore, the cans containing particles and aggregate fragments were taken on a tray. We filled these cans with caustic soda (2gr NaOH/L) since the pH of these soils was <7, again we placed the sieve holder in the working position to allow sieving until only sand particles are left on the sieve. The sieve holder was raised and placed in the leak out position, when there was no dispersion solution leaking out of the sieves anymore, the cans were removed and placed to a separate tray. The cans were placed in a convection oven at 110^0C until the water has evaporated. The weight of the materials in each can is then determined by weighing the can, plus content and subtracting the weight of the can. The stable fraction was calculated by taking the weight dry aggregates- dry sand divided by weight of (air-dried) soil- weight of dry sand*100.

3.4. Plant tissue sampling and analysis

At flowering stage, plant tissues (around 10 leaves per plot), were sampled from each treatment, prepared (oven-dried at 70°C and ground) before being analyzed for total nitrogen as described in section 3.3.4.

3.5 Growth and yield parameter measurements of climbing beans

Agronomic data were recorded along different growth stages of climbing beans. The number of nodules was recorded by uprooting and counting nodules per plant three months after planting. Yield was determined when beans were completely dried by weighing the total grain weight and using a balance. Bean grains, shoots and roots were harvested at maturity from each plot but leaving out one row on each side of the plot to minimize the edge effect. On-farm raw yield data from the previous season were also provided by the project technician and were further analyzed to enable comparison with experimental data.

3.6 Cost-benefit analysis

The cost-benefit analysis was performed based on the highest, middle and lowest yielding treatments of the experiment (4,700 kg/ha for Inoculum + Cattle manure + DAP; 3060 kg/ha for Inoculum and 2640 kg/ha for the untreated control), then extrapolated to one hectare. The expenditure was recorded for each activity or item so that it could be subtracted from the yield value, the price used was the one found at local markets for the same variety in RWF /kg grain beans. The average exchange rate considered was that of February 2012, which marked our post harvest period; 1 USD= 605 RWF.

3.7 Data management and statistical analysis

The analysis of variance (ANOVA) and t-test were done by SAS 9 software to determine whether significance exists, while means' separation was performed using LSD test. Correlation and regression were also performed to assess relationship and dependence among some variables. All the analyses were carried out at the level of significance (α=5%). Data manipulations, management and graphs were done using Microsoft Excel.

3.8 Data from farmers' fields

Data from farmers' fields were obtained from an AGRA climbing beans' project in Burera District and their raw yield data from previous season (2011A) were provided for analysis. In this case, a number of 24 farmers were considered. The experiment on farmers' fields had the same treatments and same plot size as the current research, and was designed as detailed in section 3.2.1.

CHAPTER FOUR

RESULTS AND DISCUSSION

4.0 Overview of results presentation

The results of this study are presented as follows: section 4.1 focuses on the effect of treatments on grain yields from on-station experiment and in comparison with farmers' fields. Section 4.2 presents the effect of treatments on nodulation, section 4.3 presents Nitrogen content in leaves according to treatments and in relation to grain yields and sections 4.4, 4.5, 4.6, 4.7 respectively focus on soil pH, soil organic matter, cation exchange capacity as well as cost-benefit analysis.

4.1 Effect of treatments on climbing bean grain yield

4.1.1 Yields from experimental field

The grain yields of the climbing beans as affected by various treatments are shown in Figure 5. The combined application of inoculation, FYM and DAP gave the highest mean grain bean yields of 4782 Kg/ha and was followed by Inoculation + DAP treatment (4194 Kg/ha) whereas that of the control treatment was 2640 Kg/ha, a difference of 45% from the highest yield. In comparison to the control, application of DAP, FYM and Inoculum separately increased yields by 8%, 9% and 16%, respectively.

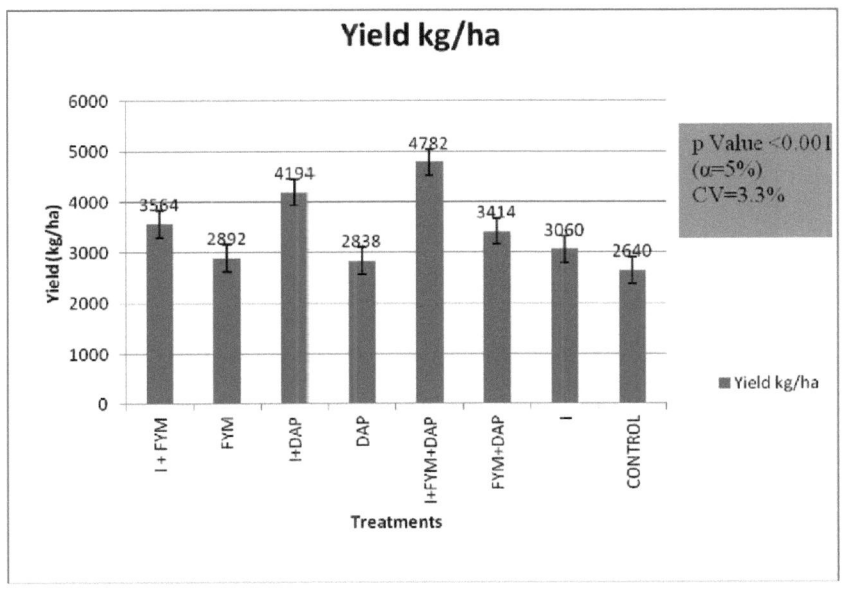

Figure 5: Effect of treatments on beans' grain yield at experimental field

There was a high statistical significant difference between Inoculation+ cattle manure+ DAP treatment and all other treatments (p<0.0001). Cattle manure and DAP applied separately did not show significant difference (p=1.000). The low response of cattle manure and DAP on climbing beans' yield could be due to low nutrient supply from cattle manure, and 50kg/ha rate of DAP (Bradley, 2008; Makungo and Odiyo, 2009). This is in line with what is found in literature, indicating that combining mineral and organic fertilizers give better results to crops (TSBF-CIAT, 2006; Verchot et al., 2007; Mbugua et al., 2007). As shown in Table 2, the overall effects of inoculation and non-inoculation significantly affected bean yield in kg/ha.

Table 2: Effect of Inoculation on climbing bean grain yield at the experimental field

Main plot	Yield kg/ha
Inoculation	3900^a
No Inoculation	2946^b
P Value	<0.0001
LSD	101
CV	3.30%

LSD= Least significant difference of means (5% level)

Means with the same letter are not significantly different.

The overall effect of inoculation was highly significant (p<0.0001) in terms of bean grain yields as given in Table 2. The main reason for that difference could be due to additional nitrogen that might have been f ixed by the legume as a result of rhizobial inoculation. Similar results have been reported by FAO, (1984) and Sessitsch *et al.*, (2002) who indicated that rhizobia are important in nitrogen fixation, thus contributing to plants N supply, which increases yield. Verchot *et al.*, (2007) also reported that nitrogen is the major limiting nutrient across areas of SSA and any additional amount of nitrogen to the soil especially through BNF, is a sustainable approach to improving nitrogen in soils for plant uptake.

4.1.2 Climbing bean grain yields from farmers' fields

Figure 6 shows yields in Kg/ha from farmers' fields as affected by different treatments.

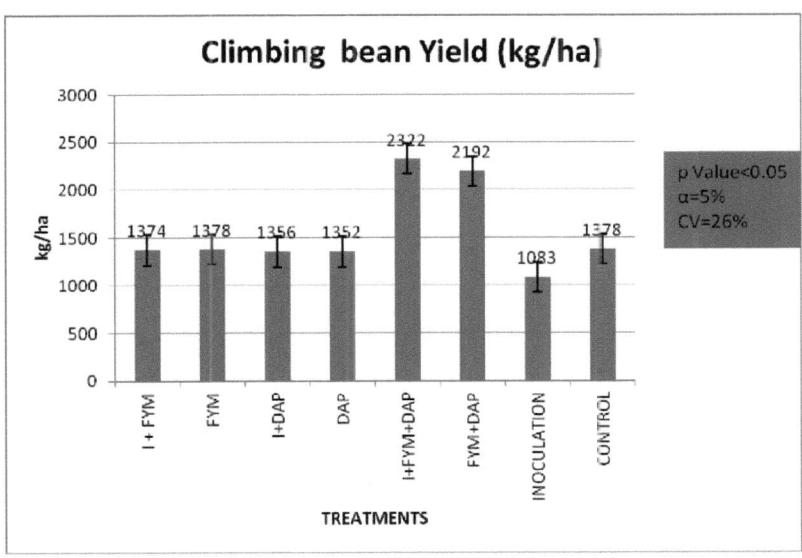

Figure 6 : Effect of treatments on beans' grain yield at farmers' fields

Like in the experimental field (Fig.5), the highest yield from farmers' fields was also obtained from the combined treatment of rhizobium inoculation, cattle manure and DAP. The treatments from the farmers' field exhibited also a similar trend in terms of yields as those from the experimental field. This shows that wherever there is a combination application of manure, bio-fertilizers and mineral fertilizers, crops yields are significantly enhanced. This observation has also been reported by Sanginga and Woomer (2009) who indicated that combining mineral and organic nutrients give better yield.

Table 3: Effect of Inoculation on bean grain yield from farmer's fields

Main plot	Yield Kg/ha
Inoculation	1534[a]
No Inoculation	1641[a]
P Value	0.681
LSD	535
CV	36%

LSD= Least significant difference of means (5% level)
Means with the same letter are not significantly different.

The main plots did not show any significant difference (p=0.681). However, non-inoculated treatments had slightly higher yields than the inoculated treatments. This may have probably been caused by poor handling of inoculants and differences in farmer management practices (RAB, 2010). RAB (2010) concluded that good crop husbandry that includes proper nutrient management, pest control and timely weeding affect crop performance.

4.1.3 Comparison of bean grain yields between the experimental field and the farmers' fields

Shown in Figure 7 are the comparisons of bean grain yields in kg/ha as obtained from the experimental field and farmers 'fields in relation to treatments. The yields obtained from the experimental field were significantly high in all the treatments as compared to those from the farmers 'fields (P (T<=t) two-tail=0.00007). The yield differences ranged between 105 and 144% with respect to maximum and minimum yields realized from the treatments. This great difference in yields could largely be explained by the management of climbing bean production used in the farmers 'fields as compared to that used in the experimental field. For example the yields obtained in the inoculation treatment on the farmers' fields were very low (1083 Kg/ha) compared to those in the experimental field (3060 Kg/ha).This explains the superiority of good management in controlled experiments as compared to farmers management. Poor handling of inputs at large, and especially, inoculation during seed – inoculant mixing, could have led to low bean yield at the farmers fields. Similar observations have been reported by other researchers (RAB, 2010 and Mugwe, 2007). Mugwe (2007) reported that there could be high variability in management among farmers, which is known to mask treatment performance whereas RAB (2010) concluded that good crop husbandry that includes proper nutrient management, pest control and timely weeding affect crop performance.

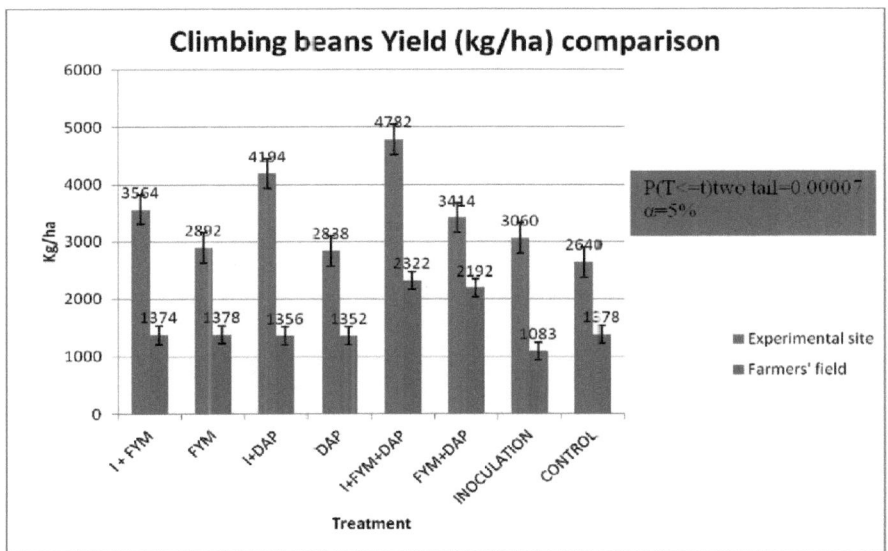

Figure 7: Comparison of bean grain yield from experimental field and farmers' fields.

4.2 Effect of treatments on nodulation

Nodulation in terms of nodule number was assessed in the 13th week after planting (in the late flowering stage, when normally maximum nodulation is attained). The parameter focused on the nodules number per plant. Table 4 shows nodule number as affected by the various treatments in the experimental field.

Table 4: Effect of treatments on the number of nodules.

Treatments	Nodule number
I + FYM	33[a]
FYM	24[b]
I+DAP	86[c]
DAP	14[d]
I+FYM+DAP	96[e]
FYM+DAP	10[d]
I	28[b]
CONTROL	14[d]
P Value	<0.0001
LSD	4.2
CV	3%

LSD= Least significant difference of means (5% level)

Means with the same letter are not significantly different.

There were high significant differences between the treatments in nodule numbers ($p<0.0001$) as indicated in Table 4. The highest number of nodules (96) was found where inoculation, cattle manure and DAP were combined followed by treatment I+DAP with 86 nodules. The lowest number of nodules was obtained in treatments FYM+DAP (10) and the control (14). This indicates that wherever inoculation, cattle manure and DAP were combined, nodulation was enhanced. This suggestion is confirmed by the work of Sessitsch *et al.*, (2002) who stated that inoculation becomes more effective in a better environment. A well-nodulated legume plant uses its own nitrogen supply that is not immediately available to weeds or companion crops which might be competing for moisture, mineral nutrients or space. The nodulated legume thus has a big advantage, particularly in a nitrogen-poor soil (FAO, 1984; Graham, 2008, Karasu et al., 2011).

Table 5: Effect of inoculation on nodulation

Main plot	Nodule number
Inoculation	60^a
No Inoculation	15^b
P Value	<0.0001
LSD	19.2
CV	3.7%

LSD= Least significant difference of means (5% level)

Means with the same letter are not significantly different.

The overall effect of inoculation on nodules formation is shown in Table 5. In terms of nodule numbers, inoculated seeds contributed four times more nodules than the non-treated seeds (60 and 15 nodules, respectively), which was highly significant (p<0.0001) in terms of main treatments. Although the difference in nodulation was highly significant between the main plots, some indigenous rhizobia strains were present in this soil and symbiotically with climbing beans could fix nitrogen. In case indigenous rhizobia strains contributed to the bean yields increase , then it could be good news for climbing bean growers because this would save them from the purchase of expensive mineral nitrogen fertilizers; whose prices are on the increase, and similarly even from the purchase of bio-fertilizers whose availability is not at reach, especially here in Rwanda (IFDC, 2009). The high nodule numbers after inoculation observed in this study has also been reported by Kala *et al* (2011) who indicated an increase in nodules number due to inoculation application.

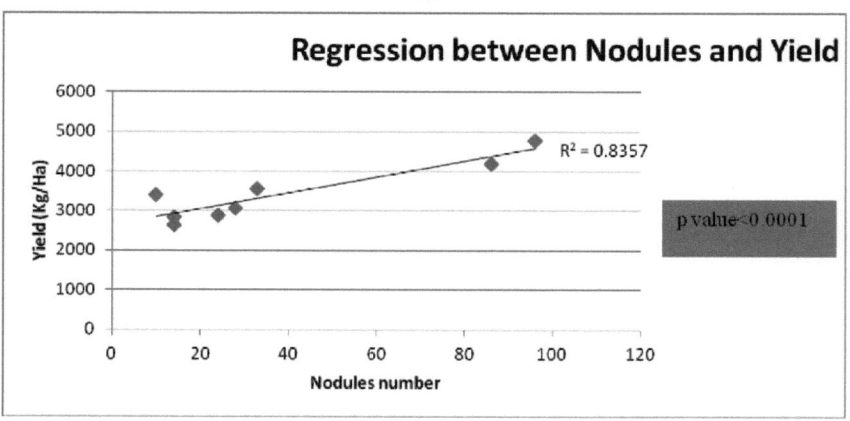

Figure 8: Regression analysis between nodule numbers and bean grain yields

A regression analysis between nodule numbers and climbing bean grain yields, as given in (Figure 8) shows that the coefficient of determination (r^2) was 0.83, an indication of a good fit. With the p value of <0.0001, this regression analysis suggests that the bean grain yields depended on nodule numbers. That is, the higher the nodule numbers the higher the bean grain yields. Other researchers have indicated similar trend, stating that if beans are not nodulated, yields often remain low, regardless of the amount of nitrogen applied (Lindemann and Glover, 2003; Rai, 2006; Mbugua *et al.*, 2007). Nodules apparently help the plant use fertilizer nitrogen efficiently (Rai, 2006).

4.3 Effect of treatments on nitrogen content in tissues

Beans' leaves were sampled and analysed in laboratory for Nitrogen content in the different treatments. Table 6 shows nitrogen content in plant tissues as influenced by the treatments.

Table 6: Effect of treatments on nitrogen content in beans' leaves.

Treatments	% Nitrogen in tissues
I + FYM	4.7^a
FYM	2.6^b
I+DAP	4.2^c
DAP	3.1^d
I+FYM+DAP	5.23^e
FYM+DAP	4.06^f
I	4.7^a
CONTROL	4.4^g
P Value	< 0.0001
LSD	0.05
CV	0.6%

LSD= Least significant difference of means (5% level)

Means with the same letter are not significantly different.

There were highly significant differences (p<0.001) between treatments. The highest nitrogen content was 5.23% and was obtained in the treatment where inoculation, cattle manure and DAP were combined whereas the lowest N contents were recorded in treatments where manure and DAP were applied separately (2.6 and 3.1%, respectively). The superiority of treatment "INOCULATION+FYM +DAP" has also been reported in this study on bean grain yields (Figs. 4 and 6), and on the nodules number of climbing bean (Table 4).This finding concurs with the findings of other researchers (Tittonel, 2008; Chivenge *et al.*, 2009) who reported the importance of combining mineral fertilizers, FYM and bio-fertilizers on legume production. In a correlation study between bean grain yields and N content in leaves, a positive correlation was found as shown in Figure 9. The analysis resulted in pearson correlation coefficient of 0.56 and a p value of 0.0037.This further emphasizes the importance of nitrogen in crop production. Nitrogen is an integral part of chlorophyll which is important in the manufacturing of carbohydrates through photosynthesis process.

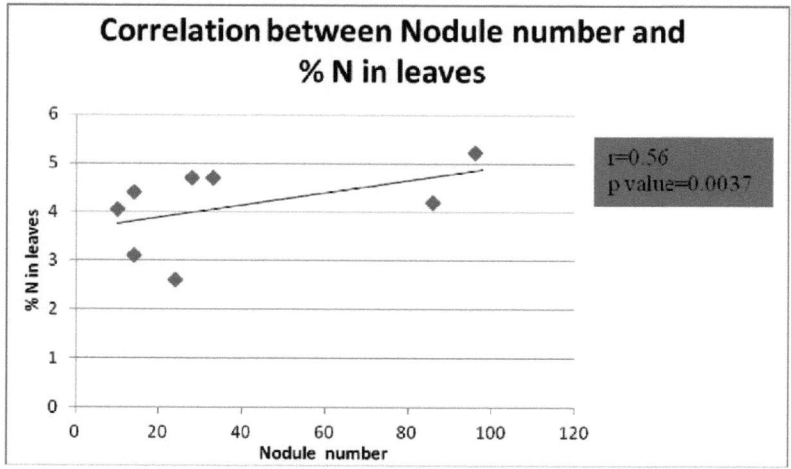

Figure 9: Correlation between N in leaves and nodule number

The effect of inoculation on nitrogen content in bean leaves was as shown on Table 7.   Inoculation of the climbing beans enhanced nitrogen content in the bean leaves (4.72%) as compared to non – inoculated beans (3.59%).

Table 7: Effect inoculation on nitrogen in beans' leaves.

Main plot	% N in tissues
Inoculation	4.72^a
No Inoculation	3.59^b
P Value	<0.0001
LSD	0.5
CV	14%

LSD= Least significant difference of means (5% level)
Means with the same letter are not significantly different.

There was a highly significant difference between the main plots (P<0.0001) and shows that inoculation enhances nitrogen fixation and its subsequent accumulation in the plant tissues. This is also the conclusion of other researchers (Sanginga and Woomer, 2009; Bambara, 2009; Silva and Uchida, 2000 and Kala et al., 2011). Bambara (2009) concluded that inoculation significantly influenced the plant total N content in different plant tissues of bean crop; which is a reflection of different levels of grain yield.

4.4 Effect of treatments on soil chemical properties

4.4.1 Effect of treatments on soil pH

Table 8 presents the active soil pH as influenced by various treatments in the climbing bean study. The results show that there was little change as compared from the control treatment (5.9).The pH levels (5.8-6.0) recorded in this experiment were favorable for climbing bean production. According to Hazelton and Murphy (2007), pH between 5.5 and 6.0 is the most acceptable range for bean production. The little change of pH as recorded in this study is an indication of a well buffered soil (mainly by high organic matter of this soil (8%)) where dynamic chemical changes are buffered and therefore take along period to show change (McCauley *et al.*, 2005).

Table 8: Effect of treatments on soil pH

Treatments	pHw
I + FYM	5.9^a
FYM	6.0^a
I+DAP	5.8^a
DAP	5.8^a
I+FYM+DAP	5.9^a
FYM+DAP	5.8^a
I	5.9^a
CONTROL	5.9^a
Grand mean	**5.9**
p Value	0.9
LSD	0.3
CV	3.3%

LSD= Least significant difference of means (5% level)
Means with the same letter are not significantly different.

4.4.2 Effect of treatments on soil organic matter

The results in Table 9 show the percentage content of organic matter in soils. The organic matter content was very high, above 7.0% and was not very much affected by experimental treatments in the one season trial (Brady and Weil, 2002). Soil organic matter is normally likely to change over a much longer period compared to just one season as suggested by McDonagh *et al.*, (2001). Moreover, in cold and wet environment, the soil organic matter tends to accumulate (Franzluebbers, 2001) which is the case in Burera District, Rwanda.

Table 9: Effect of treatments on soil organic matter

Treatments	OM %
I + FYM	7.70[a]
FYM	7.50[a]
I+DAP	7.50[a]
DAP	7.50[a]
I+FYM+DAP	7.00[a]
FYM+DAP	7.50[a]
I	7.40[a]
CONTROL	7.00[a]
Grand mean	**7.5**
p Value	0.7
LSD	0.728
CV	5.6%

LSD= Least significant difference of means (5% level)

Means with the same letter are not significantly different.

4.4.3 Effect of treatments on Cation Exchange Capacity (CEC)

Table 10 presents the soil Cation Exchange Capacity (CEC) as influenced by experimental treatments. The results show that there was little change from the control (54.17) and the experimental mean (54.48). In general, the CEC levels were fairly high, a reflection of the high organic content in this soil (Rashidi and

Seilsepour, 2008). Hartemink (2006) suggested that for most standard soil chemical properties, short term temporal variation is relatively small; therefore, the Cation Exchange Capacity of the site during one season was stable.

Table 10: Effect of treatments on soil CEC

Treatments	CEC (meq/100g soil)
I + FYM	55.27a
FYM	54.40a
I+DAP	54.33a
DAP	54.87a
I+FYM+DAP	54.67a
FYM+DAP	54.27a
I	53.93a
CONTROL	54.13a
Grand mean	54.48a
p Value	0.866
LSD	1.956
CV	2.1%

LSD= Least significant difference of means (5% level)

Means with the same letter are not significantly different.

4.5 Cost-benefit analysis

Tables 11 (a, b, c) indicate various activities with their relative cost as well as the return values of the beans yield. Based on the expenditure, the experimental input prices of different items as well as labor wage were used to calculate the cost-benefit analysis per hectare. The Cost-benefit analysis is useful in a sense that it can help the farmer in making evidence-based household decision of growing climbing beans.

In the calculation, the labor for each activity in man-days were estimated and costed as shown in the Tables 11(a,b,c).The activities included land preparation, planting and manure-fertilizers application, weeding, strings installation and staking, chemical application, and harvesting together with post-harvesting activities. Also included in the expenditure were the costs of all the inputs as indicated in the tables. In the side of returns, the grain yields (kg/ha) obtained were costed at 0.8USD/kg of bean grains (this being the average price of Gasilida bean variety used in Burera area).

For cost-benefit analysis, three scenarios were used; that is the treatment with highest yields (I+FYM+DAP = 4782 kg), the middle (Inoculation= 3060 kg) and the lowest (the Control i.e. with no inputs=2640Kg).

In the highest yielding treatment (I+FYM+DAP) scenario, a farmer is likely to get a return of around 1,330 USD per season per hectare; while in the middle and lowest yielding treatments (INOCULATION and UNTREATED CONTROL), the farmer is likely to lose 43.8 USD and 388 USD per season per hectare, respectively.

Table 11: Cost benefit analysis

a) **Highest yielding treatment (I+FYM+DAP)**

SEASON ONE					FOLLOWING SEASON
Activity	Quantity	Designation	Unit price (USD)	Total Price (USD)	Total Price (USD)
Land preparation	180	Man-day	1.2	208.3	208.3
Sowing	150	Man-day	1.2	173.6	173.6
Weeding 1	100	Man-day	1.2	115.7	115.7
Weeding 2	100	Man-day	1.2	115.7	115.7
Strings installation	200	Man-day	1.2	231.4	231.4
Spreading pesticide	50	Man-day	1.2	57.9	57.9
Harvesting +other post harvest activities	170	Man-day	1.2	196.7	196.7
Poles	6000	Number	0.2	991.7	-*
Strings	10,000	_	_	16.5	16.5
DAP	50	Kg	1.0	49.6	49.6
Biofix	1	Pack	2.5	2.5	2.5
Pesticide	40	Kg	3.3	132.2	132.2
Miscellaneous				330.6	330.6
TOTAL EXPENSES (A)	_	_	_	2,622	1,631
Value of production (B)	4,782	Kg	0.8	3,952	3,952
BENEFIT (B-A)	_	_	_	1,330	2,321

*: Poles used to support strings can last for about five years, which makes their cost a saving after season 1.

b) Middle yielding treatment (I)

SEASON ONE					FOLLOWING SEASON	
Activity	Quantity	Designation	Unit price (USD)	Total Price (USD)	Total (USD)	Price
Land preparation	180	Man-day	1.2	208.3	208.3	
Sowing	150	Man-day	1.2	173.6	173.6	
Weeding 1	100	Man-day	1.2	115.7	115.7	
Weeding 2	100	Man-day	1.2	115.7	115.7	
Strings installation	200	Man-day	1.2	231.4	231.4	
Spreading pesticide	50	Man-day	1.2	57.9	57.9	
Harvesting + other post harvest activities	170	Man-day	1.2	196.7	196.7	
Poles	6000	Number	0.2	991.7	-*	
Strings	10000	_	_	16.5	16.5	
Biofix	1	Pack	2.5	2.5	2.5	
Pesticide	40	Kg	3.3	132.2	132.2	
Miscellaneous				330.6	330.6	
TOTAL EXPENSES (A)	_	_	_	2 573	1,581	
Value of production (B)	3,060	Kg	0.8	2 529	2,529	
2BENEFIT (B-A)	_	_	_	-43.8	948	

*: Poles used to support strings can last for about five years, which makes their cost a saving after season 1.

49

c) Lowest yielding treatment (UNTREATED CONTROL)

SEASON ONE					FOLLOWING SEASON
Activity	Quantity	Designation	Unit price (USD)	Total Price (USD)	Total Price (USD)
Land preparation	180	Man-day	1.2	208.3	208.3
Sowing	150	Man-day	1.2	173.6	173.6
Weeding 1	100	Man-day	1.2	115.7	115.7
Weeding 2	100	Man-day	1.2	115.7	115.7
Strings installation	200	Man-day	1.2	231.4	231.4
Spreading pesticide	50	Man-day	1.2	57.9	57.9
Harvesting + other post harvest activities	170	Man-day	1.2	196.7	196.7
Poles	6000	Number	0.2	991.7	-*
Strings	10000	_	_	16.5	16.5
Pesticide	40	Kg	3.3	132.2	132.2
Miscellaneous				330.6	330.6
TOTAL EXPENSES (A)	_	_	_	2,570	1,579
Value of production (B)	2,640	Kg	0.8	2,182	2,182
BENEFIT (B-A)	_	_	_	-388.4	603.3

*: Poles used to support strings can last for about five years, which makes their cost a saving after season 1.

CHAPTER FIVE

CONCLUSION AND RECOMMENDATIONS

5.1 Summary and Conclusions

The aim of this study was to determine the effect of cattle manure, DAP and rhizobium inoculation on climbing beans production and soil properties changes in Burera District, Rwanda.

Bean grain yields were differently affected by treatments. The highest yields (4782kg/ha) were obtained in the treatment that received manure, DAP and Inoculant while the lowest yields were in the control (2640 kg/ha). All treatments that received inoculants did better than those without by a difference of 27%, an indication of inoculum's beneficial effect. Although yields from farmers' fields were significantly lower than those from experimental site of this study, the trend of results were similar in relation to treatments. The highest yields in both cases were found in plots which received cattle manure, DAP and rhizobium combined.

Better results of nodulation were obtained from treatments that received inoculation (60) than those without (15). Similarly, inoculated treatments exhibited higher N values (4.72 %) in plant tissues than those without inoculation which had an average of 3.59 % N. Regression analysis between nodules number and climbing beans' yield showed that the coefficient of determination R^2 was 0.83 and the p Value was 0.0001; which is an indication that yield was dependent on the number of nodules. Correlation analysis between nitrogen content in beans' leaves and yield also showed a positive correlation coefficient ($r^2=0.56$) and a p Value of 0.0037; which means that those two variables were positively related.

In general, no drastic changes were observed in soil properties as a result of inputs. In terms of cost-benefit analysis, in the highest yielding treatment (I+FYM+DAP) scenario, a farmer is likely to earn around 1,330 USD per season per hectare; while in the middle and lowest yielding treatments (Inoculation and Untreated control), the farmer is likely to lose 43.8 USD and 388 USD per season per hectare, respectively.

The findings of this study markedly indicated strong evidence that combining mineral fertilizer with cattle manure and rhizobium inoculation give the best results of climbing grain bean yields in Burera District. Also, the importance of inoculating seeds is necessary for boosting both nodulation and bean yields.

5.2 Recommendations

From the findings of this study, the following recommendations are suggested:

➤ Farmers growing climbing beans in Burera District should be encouraged to use the combination of cattle manure (20 t/ha), DAP (50 kg/ha) and rhizobium inoculation.

➤ Further research is needed to cover many seasons, and to use more rates of inputs in different agro-ecological zones so as to come up with the best rates of inputs to be recommended in Burera District and other parts of the country.

➤ Encourage participatory research with all stakeholders, and hold frequent field days to educate farmers on best approaches on inputs application and ISFM as a whole.

REFERENCES

Abiven, S., Menasseri, S., Chenu, C. 2008. The effects of organic inputs over time on soil aggregate stability. *Elsevier,* Soil Biology and Biochemistry 41:1- 12.

Africa Soil Health Consortium (ASHC). 2012. Soil health news.

Alley, M. M., and Vanlauwe, B. 2009. The role of fertilizers in Integrated Plant Nutrient Management. *IFA.* 61 pages.

Ashman, M. R., Hallett, P. D., Brookes, P. C., Allen, J. 2008. Evaluating soil Stabilization by biological processes using step-wise aggregate fractionation. *Elsevier,* Soil and Tillage Research 102:209-215.

Awad, O. A. and Eltahir, A. O., 2011. Effect of intercropping, phosphorus fertilization and rhizobium inoculation on the growth and nodulation of some leguminous and cereal forages. *Agriculture and Biology Journal of North America,* 2(1): 109-124.

Azeez, J. O. and Van Averbeke W. 2010. Nitrogen mineralization potential of three animal manures applied on a sandy clay loam soil. Tshwane University of Technology. *Elsevier,* Bioresource Technology 101:5645–5651.

Bambara, K. S. 2009. Effect of Rhizobium inoculation, molybdenum and lime on the growth and Nitrogen fixation in *P. vulgaris.* L. Cape Peninsula University of Technology. *Theses and Dissertations.* 106 pages.

Bekunda, M., Sanginga, N. and Woomer P.L. 2010. Restoring soil fertility in Sub-Sahara Africa. *Elsevier,* Advances in Agronomy, Volume 108.

Bot, A. and Benites, J. 2005. Importance of soil organic matter. Key to drought resistant soil and sustained food and production. *FAO Soils Bulletin*, Rome. 95 pages.

Bradley, L. A. 2008. Manure management for small and hobby farms. *Northeast Recycling Council, Inc.*, Manure Management Handbook. 27 pages.

Brady, N. and R. Weil. 2002. The Nature and Properties of Soils, 13th Edition. Prentice Hall, Upper Saddle River, New Jersey. 960 pages.

Bray, R.H., and Kurtz, L.T. (1945). Determination of total organic and available phosphorus in soils. *Soil Science* 59:39-45.

Burdass, D. and Hurst, J. 2002. Rhizobium, root nodules and Nitrogen fixation. *Society for general microbiology*. A post - 16 resource.

Carter, M.R. and Gregorich, E.G. 2006. Soil sampling and methods of analysis, Second Edition. *Taylor and Francis Group*, Canadian society of soil science. 198 pages.

Chapman, H.D. 1965. Cation-exchange capacity. In: C. A. Black (ed.) Methods of soil analysis - Chemical and microbiological properties. *Agronomy* 9: 891-901.

Chivenge, P., Vanlauwe, B., Gentile, R., Wangechi, H., Mugendi, D., van Kessel, C. and J. Six, J. 2009. Organic and Mineral Input Management to Enhance Crop Productivity in Central Kenya. *Agronomy journal,* Vol 101, issue 5, 101:1266–1275

Czymmek, K., Ketterings, Q., Van H. E., DeGloria, S. and Albrecht, G. 2005. The New York Nitrate leaching Index. *Agronomy Fact Sheet Series,* Fact Sheet 11. Cornell University Cooperative Extension.

Ebesu, R. 2004. Home garden beans. College of Tropical Agriculture and Human Resource. Department of Plant and Environmental Protection Sciences. Home Garden Vegetable-8. University of Hawaii.

Feed the Future (FTF). 2011. Global food security research strategy. A U.S Government initiative. United Stated of America. 50 pages.

Food and Agriculture Organization (FAO). 2000. Fertilizer Requirements in 2015 and 2030. Rome. 37 pages.

Food and Agriculture Organization (FAO). 1984. Legumes inoculants and their use. FAO Pocket Manual. 75 pages.

Franzluebbers A.J. 2001. Soil organic matter stratification ratio as an indicator of soil quality. *Elsevier*, Soil and Tillage Research 66:95–106.

Fujishige, N. A., Michelle, R. L., De Hoff Peter L., Whitelegge, J. P., Faull Kym F. and Hirsch, A. M. 2008. Rhizobium common nod genes are required for biofilm formation. Department of molecular, cell and developmental biology, Molecular biology Institute, Pasarow Mass Spectrometry Laboratory and Semel Institute, University of California.

Graham, P.H. (2008). Ecology of the root-nodule bacteria of legumes. Chapter (2), p. 23-43. In Dilworth, M. J, James, E. K, Sprent, J. I and Newton, W.E. (eds.) (2008). Nitrogen-fixing Leguminous Symbioses. Nitrogen Fixation: Orgins, Application. Volume 7, *Springer*. 402 pages.

Hanson, B. and Orloff, S. 1998. Measuring soil moisuture. Irrigation program. University of California. Engeneer Handbook. 34 pages.

Hartemink, A. E. 2006. Assessing soil fertility decline in the tropics using soil chemical data. *Elsevier,* Advances in Agronomy, volume 89. 47

pages.

Hazelton, P. and Murphy, B. 2007. Interpreting soil test results. What do all the numbers mean? CSIRO Publishing. Australia. 169 p.

Herrick, J.E., W.G. Whitford, A.G. de Soyza, J.W. Van Zee, K.M. Havstad, C.A. Seybold, and M. Walton. 2001. Field soil aggregate stability kit for soil quality and rangeland health evaluations. *Elsevier,* Catena 44:27-35.

Hirsch, A. M. 2009. Brief history of the discovery of Nitrogen-fixing Organisms. *Nature,* Vol. 411 21.

Hutchins, A. M. 2011. Bean briefs.*US Dry Bean Council.* University of Colorado.

Institut des Sciences Agronomiques du Rwanda (ISAR). 2009. Bean Program. Rwanda.

International Fertilizer Development Center (IFDC). 2009. An Action for developing agricultural inputs markets in Rwanda. Alabama. USA. 79 pages.

International Fertilizer Development Center (IFDC). 2002. Improving Agricultural input supply systems in Sub-Saharan Africa: A review of literature. Alabama. USA. 48 pages.

International Fertilizer Industry Association (IFIA). 2000. Mineral fertilizer use and the environment. Revised edition. Paris. 53 pages.

Kala, T. C., Christi, M. R. and Bai, R. N. 2011. Effect of Rhizobium inoculation on the growth and yield of Horsegram (*Dolichos Biflorus*). *Plant Archives* Vol. 11 No. 1, 97-99.

Karasu, A., Oz, M. and Dogan, R. 2011. The effect of bacterial inoculation and different nitrogen doses on yield and yield components of some dwarf dry bean cultivars (*Phaseolus vulgaris*). *Bulgarian Journal of Agricultural Science.* 17: 296-305.

Kelly, J. D., Peralta, E. and Butare, L. 2012. Combining conventional, molecular and farmer participatory breeding approaches to improve Andean Beans for resistance to biotic and abiotic stresses in Ecuador and Rwanda. Pulse CRSP.

Kemper, W.D., Koch, E.J., 1966. Aggregate stability of soils from western United States and Canada.USDA-ARS Tech. Bull., vol. 1355. U.S. Govt. Print. Office, Washington, DC

Ketterings, Q., Reid, S., and Rao, R. 2007. Cation exchange capacity. Agronomy Fact Sheet Series. Fact Sheet 22. Cornell University Cooperative extension.

Lindemann, W.C. and Glover, C.R. 2003. Nitrogen fixation by legumes. Cooperative Extension Service, Guide A-130. New Mexico State University.

Mafongoya, P. L., Bationo, A., Kihara, J. and Waswa, B. S. 2006. Appropriate technologies to replenish soil fertility in southern Africa. *Springer,* Nutr. Cycl. Agroecosyst 76:137–151.

Makungo, R. and Odiyo, J.O. 2011. Determination of steady state infiltration rates for different soil types in selected areas of Thulamela Municipality. South Africa. 15th SANCIAHS National Hydrology Symposium proceedings, Rhodes University, Grahamstown.

Mbugua, G.W., Wachiuri, S.M., Karoga, J. and Kimamira, J. 2007. Effects of commercial rhizobium strain inoculants and triple superphosphate fertilizer on yield of new dry bean lines in Central Kenya. *KARI publications*. Kenya.

McCauley, A., Jones, C. and Jacobsen, J. 2005. Basic soil properties. Module 1. Montana State University. 12 pages.

McDonagh, J.F., Thomsen, T. B. and Magid, J. 2001. Soil organic matter decline and compositional change associated with cereal cropping in southern Tanzania. *John Wiley & Sons, Ltd,* Land Degrad. Develop. 12: 13±26.

Ministry of Agriculture and Animal Resources (MINAGRI). 2004. National Agriculture Policy. Kigali. Rwanda.

Ministry of Agriculture and Animal Resources (MINAGRI). 2011. Crop assessment-Season 2011B. Kigali, Rwanda. 27 pages

Ministry of Agriculture and Animal Resources (MINAGRI). 2011. Strategies for sustainable crop intensification in Rwanda. Kigali, Rwanda. 59 pages.

Ministry of Finance and Economic planning (MINECOFIN). 2002. Rwanda Vision 2020. Kigali, Rwanda. 29 pages.

Mugwe, J. N. 2007. An evaluation of integrated soil fertility management practices in Meru South District, Kenya. PhD Thesis, *Kenyatta University*. 191 pages

National Institute of Statistics of Rwanda (NISR). 2008. District Baseline Survey. Burera District. Kigali, Rwanda. 84 pages

National Institute of Statistics of Rwanda (NISR). 2012. 2012 Population and Housing Census. Provisional results. Kigali, Rwanda. 52 pages.

Okalebo, J. R. 2009. Recognizing the constraint of soil fertility depletion and technologies to reverse it in Kenyan Agriculture. Inaguration presentation, Moi University. 61 pages.

Okalebo, J.R., Gathua K.W. and Woomer, P.L. 2002. Laboratory Methods of Soil and Plant Analysis: A Working Manual. Tropical soil Biology and Fertility Programme, Nairobi. 128 pages.

Okalebo, J. R., Othieno, C. O., Woomer, P. L., Karanja, N. K., Semoka, J. R. M.,Bekunda, M. A., Mugendi, D. N., Muasya, R. M., Bationo, A. and Mukhwana, E.J. 2006. Available technologies to replenish soil fertility in East Africa. Springer journals. Vol 76.

Pauwels, J. M., Van Ranst, E., Verloo, O. and Mvondo, Z. A. 1992. Manuel de Laboratoire de pédologie. Méthodes d'analyses de sols et de plantes, Equipement, gestion de stocks de verrerie et de produits chimiques. *Publications agricoles-28*. Bruxelles, Royaume de Belgique. Pp 25-32. 658 pages.

Rai, M. K. 2006. Handbook of microbial biofertilizers. *Haworth Press, Inc.,* Binghamton, New York. 42 pages.

Rascio, N. and La Rocca, N. 2008. Ecological Processes/Biological Nitrogen Fixation. *Elsevier,* vol 412.

Rashidi, M. and Seilsepour, M. 2008. Modeling of soil cation exchange capacity based on soil carbon. Asian Research Publishing Network. *Journal of Agricultural and biological sciences.* Volume 3. N°.4, 45 pages.

Rwanda Agriculture Board (RAB). 2010. Integrated Bean crop management for increased productivity in Rwanda. *Extension guide*. Rwanda.

Saikia, S. P. and Jain V. 2007. Biological nitrogen fixation with non-legumes: An achievable target or a dogma? Current science, vol. 92, no. 3.

Sanchez, P. A. 2002. Soil fertility and hunger in Africa. Policy forum. *Science,* Vol. 295.

Sanginga, N. and Woomer, P. L. 2009. Integrated Soil Fertility Management in Africa: Principles, Practices and Developmental Process. Tropical Soil Biology and Fertility. International Center for Tropical Agriculture. Nairobi. 263 pages.

Sessitsch, A., Howieson, J.G., Perret, X., Antoun, H. and Martinez-Romero, E. 2002. Advances in Rhizobium Research. Critical reviews in plant science. 21:323–378.

Shoji, S., Nanzyo, M., and Dahlgren, R. A. 1993. Volcanic ash soils- Genesis, Properties and Utilization. 1st Edition, *Elsevier,* vol 21.

Silva, J. A. and Uchida, R. 2000. Biological Nitrogen Fixation. Nature's Partnership for Sustainable Agricultural Production. Chapter 13. Technical bulletin. College of Tropical Agriculture and Human Resources, University of Hawaii at Manoa.

Tang, J., Mo, Y., Zhang and J., Zhang, R. 2011. Influence of biological aggregating agents associated with microbial population on soil aggregate stability. Applied Soil Ecology 47:153–159.

Tittonell, P., Corbeels, M., Van Wijk, B. M. T., Vanlauwe, B. and Giller, K. E. 2008. Combining Organic and Mineral Fertilizers for Integrated Soil Fertility Management in Smallholder Farming Systems of Kenya: Explorations Using the Crop-Soil Model FIELD. Agronomy journal. Vol 100, issue 5.

Tropical Soil Biology and Fertility Institute of the International Center for

Tropical Agriculture (TSBF-CIAT). 2006. Integrated soil fertility management in the tropics.TSBF-CIAT's Achievements and Reflexions 2002-2005.   Colombia.101 pages.

Uphoff, N., Ball, A., Fernandes, E., Herren, H., Husscn, O., Laing, M., Palm, C., Pretty,J., Sanchez, P., Sanginga and N., Thies, J. 2006. Biological Approaches to Sustainable Soil Systems. CRS Press. Chapter 5. Pp 60-76. 784 pages.

Verchot, L.V., Place F., Shepherd K.D and Jama B. 2007. Science and Technological Innovations for Improving Soil Fertility and Management in Africa. A Working Paper number 30. A report for the NEPAD Science and

Technology Forum. World Agroforestry Center. Working paper n° 31. Nairobi, Kenya. Pp. 102.

Wallace, M. B., and Knausenberger, W. I. 1997. Inorganic fertilizer use in Africa: Environmental and economic dimensions. Applied Research, Technical Assistance and Training. Winrock International Environmental Alliance Arington, Virginia. USA. 63 pages.

World Food Programme (WFP). 2009. Comprehensive Food Security and Vulnerability Analysis & Nutrition Survey. 108 pages.

http://www.burera.gov.rw/index.php?id=368 consulted on June, 23rd, 2013.

APPENDICES

Appendix 1: Treatments randomization in plots.

PLOT NUMBER	REPLICATION	TREATMENT
1	1	INOC
2	1	FYM
3	1	INOC+FYM+DAP
4	1	INOC+DAP
5	1	FYM+DAP
6	1	DAP
7	1	INOC+FYM
8	1	CONTROL
1	2	INOC+FYM+DAP
2	2	INOC
3	2	DAP
4	2	CONTROL
5	2	FYM+DAP
6	2	INOC+FYM
7	2	INOC+DAP
8	2	FYM
1	3	DAP
2	3	INOC+FYM
3	3	CONTROL
4	3	INOC+DAP
5	3	INOC
6	3	FYM
7	3	INOC+FYM+DAP
8	3	FYM+DAP

Appendix 2: USDA textural triangle

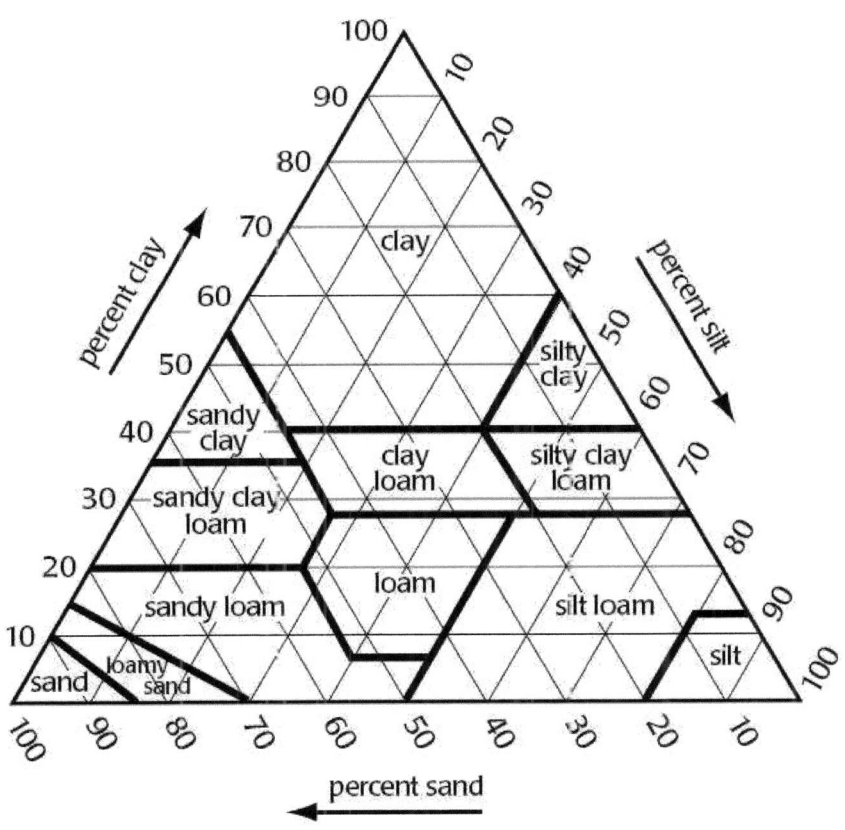

Appendix 3: EXPERIMENTAL SITE LOCATION MAP

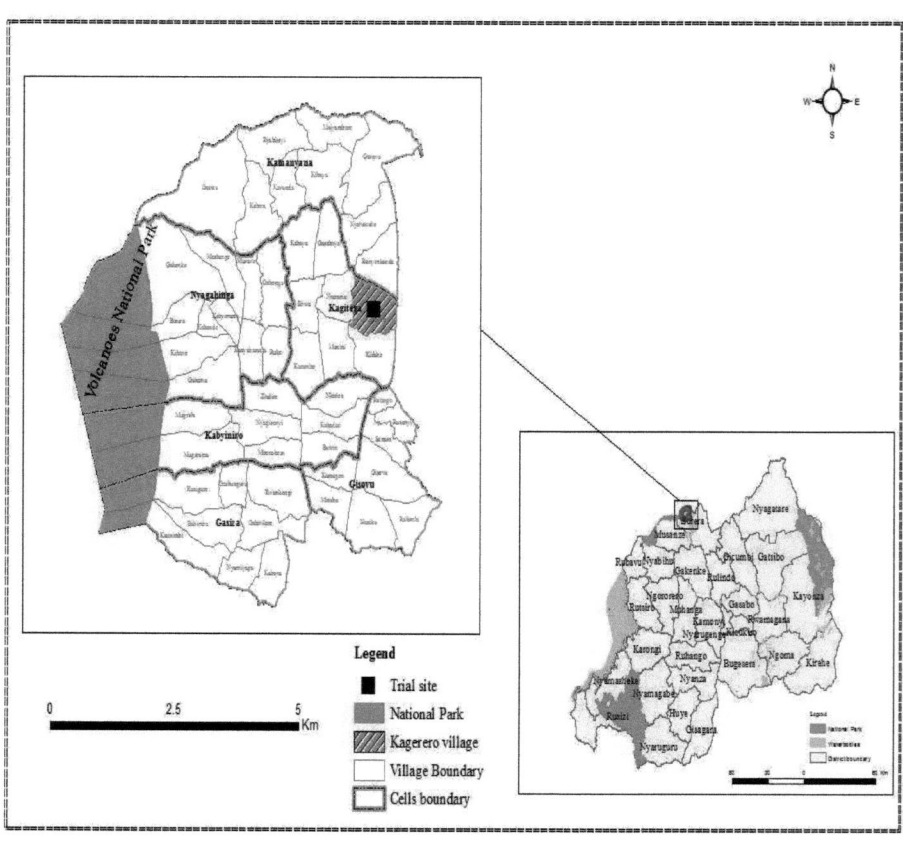

Appendix 4: Formulas used in soil and plant leaves laboratory analysis

(i) %N= (a-b)*0.2*v*100/1000*w*al

Where;

a: Volume of the titre HCl for the blank

b: Volume of the titre HCl for the sample

v: Final volume of the digestion

w: weight of the sample taken

al: aliquot of the solution taken for analysis

(ii) Texture

Corrected hydrometer readings

a) Corrected hydrometer reading at 40 seconds (PSH40COR):

(PSH40SAM - PSH40BLK) + [(PST40 - 20) 0.36]

b) Corrected hydrometer reading at 2 hours (PSH2HCOR):

(PSH2HSAM - PSH2HBLK) + [(PST2H - 20) 0.36]

Where;   PSH40SAM = Hydrometer reading at 40 seconds for sample

PSH40BLK = Hydrometer reading at 40 seconds for blank

PST40 = Temperature at 40 seconds

PSH2HSAM = Hydrometer reading at 2 hours for sample

PSH2HBLK = Hydrometer reading at 2 hours for blank

PST2H = Temperature at 2 hours

2.   Percent clay (CLAY) = (PSH2HCOR) 100

PSSLWT

Where; PSSLWT = Weight of air dry soil (g)

3. Percent sand (SAND) =100 - [(PSH40COR) 100]
                                        PSSLWT

4. Percent silt (SILT) = 100 - SAND - CLAY

(iii) %C= (a-b)*0.1/w

Where;

a: Concentration of chromic ions (Cr^{3+}) in the sample

b: Concentration of chromic ions (Cr^{3+}) in the blank

w: Weight of the sample

(iv) Available N

NO_3^-(mg/kg) : 11.2(V2-V0')
                      (1-Hf)

NH_4^+(mg/kg): 11.2(V1-V0)
                     (1-Hf)

Where;

Hf: Moisture content of the sample

V0 : Volume of HCl 0.01 N used for ammonium titration in the sample

V0': Volume of HCl 0.01 N used for ammonium titration in the blank

V1: Volume of HCl 0.01 N used for ammonium titration

V2: Volume of HCl 0.01 N used for nitrates titration

(v) Available P (ppm) = AAS reading X dilution factor.

Dilution factor used: 43.75

(vi) K (ppm): AAS Reading x dilution factor.

Dilution factor used: 200

(vii) CEC : (ml titr.-
ml blanco) x 4

Appendix 5: ANOVA tables

Table 1: Effect of Cattle manure, DAP and rhizobium inoculation on climbing beans yield

Treatments	Yield kg/ha
I + FYM	3564^a
FYM	2892^b
I+DAP	4194^c
DAP	2838^b
I+FYM+DAP	4782^d
FYM+DAP	3414^e
I	3060^f
CONTROL	2640^g
P Value	<0.0001
LSD	134
CV	3.30%

LSD= Least significant difference of means (5% level)
Means with the same letter are not significantly different.

Table 2: Effect of Cattle manure, DAP and rhizobium inoculation on climbing beans yield from farmer's field

Treatments	Yield in kg/ha
I + FYM	1374^a
FYM	1378^a
I+DAP	1356^a
DAP	1352^a
I+FYM+DAP	2322^b
FYM+DAP	2192^b
I	1083^a
CONTROL	1378^a
P Value	<0.0001
LSD	723
CV	26%

LSD= Least significant difference of means (5% level)
Means with the same letter are not significantly different.

Printed by Books on Demand GmbH, Norderstedt / Germany